Strengthening Government Laboratory Science and Technology Programs:

Some Thoughts for the Department of Homeland Security

Samuel Musa, Richard Chait, Vincent Russo, and Donna Back

Center for Technology and National Security Policy
National Defense University

July 2011

1

The views expressed in this article are those of the authors and do not reflect the official policy or position of the National Defense University, the Department of Defense or the U.S. Government. All information and sources for this paper were drawn from unclassified materials.

Samuel Musa is a Senior Research Fellow at the Center for Technology and National Security Policy (CTNSP). He was previously Associate Vice President for Strategic Initiatives and Professor of Electrical and Computer Engineering at Northwestern University. Dr. Musa received his PhD in Applied Physics from Harvard University and a BS in Electrical Engineering and a BA from Rutgers University.

Richard Chait is a Distinguished Research Fellow at CTNSP. He was previously Chief Scientist, Army Materiel Command, and Director, Army Research and Laboratory Management. Dr. Chait received his PhD in Solid State Science from Syracuse University and a BS degree from Rensselaer Polytechnic Institute.

Vincent J. Russo is the former Executive Director of the U.S. Air Force Aeronautical Systems Center. Dr. Russo received his PhD in Metallurgical Engineering from the Ohio State University, a MS from the Air Force Institute of Technology, and BS from University of Rochester.

Donna J. Back is the Chief Operating Officer and Chief Financial Officer of Growing Splendid Leaders, LLC. She serves as a leadership instructor for Wright State University's Raj Soin College of Business, University of Dayton's College of Engineering, and the Ohio State University's College of Engineering. She received her MBA from Wright State University and a BS from Miami University, Ohio.

Acknowledgments

The authors gratefully acknowledge the support from Starnes Walker, former Director of Research, Rolf Dietrich, Acting Deputy Undersecretary, and Capt. David Newton, Office of University Programs, Science and Technology Directorate, Department of Homeland Security. Also, we are pleased to acknowledge Timothy Coffey, John Lyons, and William Berry for their contributions during the early stages of the study, and Samuel Bendett, Research Associate at CTNSP, who assisted the team and provided editing of the paper.

Defense & Technology Papers are published by the Center for Technology and National Security Policy, National Defense University, Fort Lesley J. McNair, Washington, DC. CTNSP publications are available at http://www.ndu.edu/ctnsp/publications.html.

Table of Contents

Introduction

Members of the Center for Technology and National Security Policy (CTNSP) at the National Defense University in Washington, DC, examined various management practices and implications of laboratory administration for the U.S. Department of Homeland Security (DHS). This paper summarizes the second phase of a research and analysis project that stemmed from the initial work which provided DHS Science and Technology (S&T) leadership with examples of practical approaches to risk-informed decisionmaking and metrics for program and project selection.[1] The second phase was undertaken to provide additional relevant information to DHS as it seeks to strengthen its laboratory programs.

This paper is in three parts. Part One summarizes the CTNSP Team's (hereafter referred to as the Team) key lessons learned based on their experiences leading Department of Defense (DOD) laboratories. A list of the Team members is provided in Appendix A. In Part One, the Team developed responses to several questions posed by DHS leadership. To develop Part One, the Team compiled a study on the management and organization of DOD laboratories and documented lessons learned from their personal experiences leading laboratories in the Army, Navy, and Air Force. Additionally, the Team conducted research into existing methods and practices outlined in journals, periodicals, and publications dealing with the topic of managing laboratories. Part Two of this paper contains a synopsis of the factors considered by the Team to be important to the development of an S&T workforce. In Part Three, a summary and recommendations are provided.

To supplement the text contained in the body of this report, several Appendices are provided. Appendix B presents other views of important characteristics of successful laboratories. In Appendix C, a summary is provided of the work of a Team member (Russo) and his co-author (D. J. Back), on leading technical organizations. Finally, Appendix D contains a summary description of the various DHS laboratories.

The concepts presented here summarize the results of reviews of relevant academic research, exposure to broad executive training, consultation with government leaders, and extensive experience by the authors in leading a variety of S&T organizations. It is hoped that understanding these concepts will help S&T organizations within the DHS and elsewhere to fulfill their vision, achieve their mission, and produce high-quality laboratories with a dedicated, productive and motivated workforce.

---

[1] *Risk-Informed Decisionmaking for Science and Technology*, Defense and Technology Paper 76, Washington, DC, September 2010.

PART ONE

Best Practices—DOD Laboratory Leaders

In order to discuss the practice of leading government laboratories, the CTNSP Team first outlined the best management practices for leading DOD laboratories. These practices were drawn from the Team members' personal experiences, as well as from literature on the subject. This resulted in a framework for consideration as DHS works to overcome the challenges faced in their organizations. Clearly, there are several differences between what makes a successful DOD laboratory, and what makes a successful DHS laboratory. But there are also be several similarities the authors believe are fundamental to the operations of good government laboratories.

In order to set the stage for this chapter, it is necessary to introduce the organization of the DHS Science and Technology Directorate. The organizational chart is shown below:

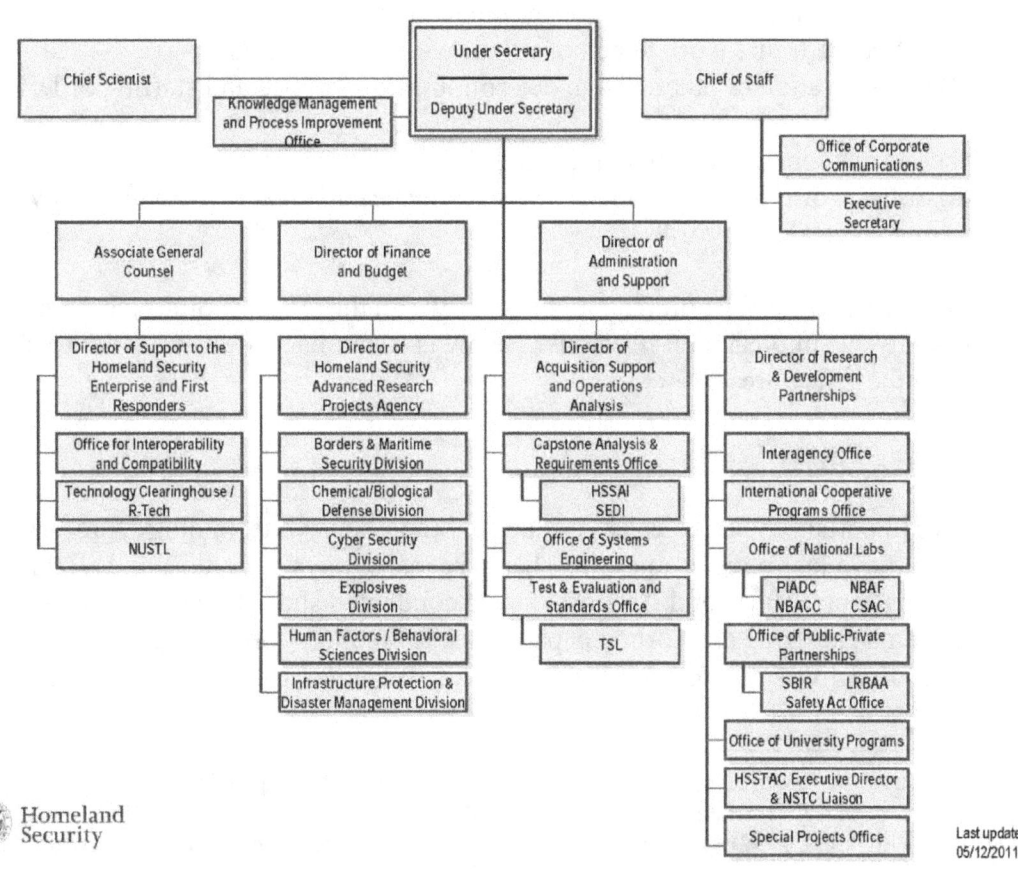

The Directorate is now organized to better serve its customers in DHS and the state and local first responder community. It consists of four groups. The Homeland Security Advanced Research Projects Agency (HSARPA) now serves as the home to all technical units within the directorate. The second group is the Homeland Security Enterprise and First Responders, which serves as the lead for supplying solutions to state and local government agencies. One DHS Laboratory, National Urban Security Technology Laboratory (NUSTL), is part of this group. The Acquisition Support and Operations Analysis group devotes its activities to assisting operational DHS components with acquisitions. One Laboratory, Transportation Security Laboratory (TSL), reports to this group. The Research and Development Partnerships group focuses on resources DHS can leverage when conducting research and development. Three of the DHS Laboratories are part of this group.

 The former Deputy Director of Research at DHS Science & Technology Directorate (now Director for Knowledge Management and Process Improvement) requested the Team to investigate the best practices in DOD laboratories in five areas:
> 1) Forecast future work and relationship to DHS Labs;
> 2) Influence future work to be consistent with requirements;
> 3) Consideration of other labs in determining capabilities and staffing of DHS labs;
> 4) Information about competitors; and
> 5) Sources of information and practices in generating resources such as physical infrastructure and staffing profile.

All of the areas addressed an ultimate theme—how can the DHS laboratories be led best to successfully accomplish their missions? Below is a summary of the CTNSP Team's thoughts on these five areas.

1. Forecasting Future Work

It is always important to periodically conduct a comprehensive technology forecast in areas relevant to a laboratory's mission. There are many ways to do this, and all three Departments (Army, Navy, and Air Force) have conducted such studies[2]. With today's web-based data gathering and sorting capabilities, technology forecasting can be a relatively short activity compared to past practices. Whatever technique is used to

---

[2] Board on Army Science and technology, Commission on Engineering and technical systems, National Research Council, STAR 21-Strategic Technologies for the Army Twenty-First Century (Washington, DC: National Academy Press, 1992); Naval Studies Board, National Research Council, Technology for the United States Navy and Marine Corps, 2000–2035 (Washington, DC: National Academy Press, 1997); Air Force Scientific Advisory Board, *New World Vistas, Air and Space Power for the 21st Century* (Washington, DC: Department of the Air Force, 1995).

accomplish the forecasting, it is important to insure the ultimate user of the technology has the ability to influence the resulting priorities. The user should be engaged in the process from day one. Additionally, there should be strong connections to industry and universities during the forecasting process. Industry, through its independent research and development (IR&D) programs, can provide ideas, conduct early research, and develop proof of concepts. Their involvement in the forecasting activities also helps prepare them for subsequent contractual activities. The relationship with the universities, (in particular for DHS their University Centers of Excellence (CoEs)), should be long-lasting to assure continuity and connectivity. It is important to involve academic experts in the forecasting efforts. Mechanisms to incorporate this expertise via tools such as Internal Governmental Personal Act (IPAs), on a limited-time basis, have been proven to be effective. These IPAs can supplement the talent of the laboratory and industry workforces. Finally, the laboratory staff, if they are "card-carrying" members in their fields of science and technology and are held in high regard, can be an excellent resource for forecasting. Without highly knowledgeable staffs who are members of the scientific community in good standing, successful forecasting will always be in doubt.

2. Influence Future Work to be Consistent with Requirements

Each of the DOD Military Departments has an organization that tracks or monitors technology advances: Army Research Office, Office of Naval Research, and Air Force Office of Scientific Research. Part of their function is to interface with the DOD Laboratory counterparts to assure continuity from basic research to exploratory development and to be a source for identifying and funding the early stages of a new technology. Likewise, each Service has a formalized method of identifying the requirements from their users. With inputs from their forecasting activities, their basic research organizations, their user identified requirements, and their own staffs, DOD laboratory managers then identify work to be performed within allocated budgets.[3] The managers are then responsible to advocate their work to appropriate levels for approval. Key to accomplishing these tasks is a laboratory staff that is skilled in these various activities.

3. Consideration of Other Laboratories in Determining Capabilities and Staffing

DHS, like any other government agency, should seek ways to "tap into" the broad and robust capabilities in the government laboratory system (National Institutes of Health, Commerce, DOD, NASA, etc.) for technical authority. These capabilities are well established and can be readily accessed. Experience within the DOD has shown the different missions of each laboratory often result in work being performed at one lab can be leveraged at another lab. Additionally, joint funding of common use technology leverages limited resources and adds to the productivity of laboratories. Important to any involvement with the DOD labs is the consideration of multi-year commitments in order to stimulate cooperation. In the coming years, defense technologies should become increasingly relevant to DHS. This provides an opportunity for DHS to form real partnerships with DOD and with the DOD laboratories.

---

[3] See NDU paper "Risk-Informed Decisionmaking for Science and Technology," September 2010.

In addition to the classical government-owned/government-operated (GOGO) and the government-owned/contractor-operated (GOCO) laboratory models, another effective model for the DHS laboratories to consider is the government-owned/contractor-assisted (GOCA) laboratory. In such a laboratory, typically 50–60 percent of the work is performed by government employees and 40–50 percent is performed by contractors. This concept provides the stability necessary via the government workforce while at the same time providing the flexibility provided via the contracted workforce. The "contractor" workforce in this case could consist of IPAs, visiting scientists, and post docs in addition to a classic contract with a research organization.

4. Information about Competitors

The DOD laboratories usually conduct peer reviews by externally recognized experts, which generally provide good information about the competition in other labs within and outside DOD. Leveraging knowhow of competitor/peer laboratories instead of engaging in unnecessary duplication of effort should be an important goal of any government laboratory. An example of efforts to avoid unnecessary duplication is DOD Project Reliance[4]. Project Reliance was formed to provide a mechanism for sharing information among the DOD labs that allowed them to better coordinate their efforts.

Another approach is to organize a laboratory by discipline or multidiscipline versus a project basis. This is widely practiced in the Army and Navy Labs. In the case of DHS Laboratories, this approach may be difficult since they are charged to advance the science and technology in support of relatively near term products. At a higher level, Office of Science and Technology Policy (OSTP) tries to provide interagency coordination. Finally, a good and well-connected staff should have a broad-based understanding of the capabilities of competitors. They should be in a position to provide relevant information about the future work of competitors.

5. Sources of Information and Practices in Generating Resources

Regarding funding opportunities, good advocacy processes by the senior leaders of a laboratory are essential. Additionally, the stature and connectivity of the staff with the ultimate users of the technology are critical. Processes that routinely engage users to help establish funding priorities are valuable to help create a community of supporters who can assist with defending budget requests.

A suggested approach to protecting and growing the budget is to conduct periodic meetings with congressional staffers. In the DOD Laboratories, this practice is usually accomplished by having the appropriate members of the Service's legislative staff coordinate such activities.

---

[4] http://www.fas.org/irp/offdocs/pdd5status-b html

8

With respect to physical infrastructure, one approach suggested is to allocate a fixed percentage of a laboratory's annual budget to upgrading equipment. Outstanding laboratories need outstanding equipment, and a clearly identified source for funding equipment is important.

In addition to the five areas discussed above, the Team also responded to the following requested areas of evaluation identified during discussions with DHS leadership.

A. Balance of short term technology solutions versus long-term research.

Based on their experiences leading DOD laboratories, the Team defined the duration of research as follows: near-term for 1-2 years; mid-term for 3–5 years; and long term for 5+ years. The Team suggested the following distribution for research laboratories: approximately 55 percent of the research should be near term, 30 percent mid-term, and 15 percent long term[5]. These numbers are proposed for a full-spectrum laboratory; for development and engineering laboratories, the distribution may be different.

B. Interagency and interdepartmental coordination of basic research.

The Team recommended the following: the driver for interagency coordination must be at the OSTP level in order to be successful, and there are specific committees that handle such coordination. They are the OSTP Federal Interagency Coordination Research Group and the OSTP Committee for Homeland and National Security. The Team also suggested a peer review process independent from and external to DHS be considered. Furthermore, DHS should leverage other organizations (NIH, Service S&T research offices etc.) for longer-term research.

C. Quick response to new urgent requirements.

The Team recommended DHS identify a dedicated group of individuals throughout the country whose expertise can immediately be brought to any problem or emergency. The concepts of "One phone call away from the Country's best" and "Bags packed and ready to go" should be followed. In order for this concept to be successful, it is necessary to have the appropriate contracts/agreements in place. Examples include such existing organizations as the Army Rapid Equipping Force (REF)[6] and the Joint Improvised Explosive Device Defeat Organization (JIEDDO)[7].

D. Cost of doing business in a Laboratory

---

[5] John Lyons and Richard Chait, "Assessing the Health of Army Laboratories, Funding the Basic Research and Laboratory Capital Equipment," Defense & Technology Paper, Center for Technology and National Security Policy, National Defense University, September 2010.
[6] http://www.ref.army.mil/portal/
[7] https://www.jieddo.dod.mil/index.aspx

There are two components of interest here. One is the direct expense that includes salaries and benefits. The other is overhead. Salaries and benefits are comparable from one lab to another in the public sector and not too different in the private sector. Higher salaries in the private sector are offset by the generally better benefits in the Federal government. Overhead is often cited as a differentiating factor. The problem with this is that there is no standard definition of overhead or of the means of expressing it. In general, the costs of investing in a government lab are the same or slightly less than that of the private sector.

The controlling factor is the competitiveness of the lab (excellence of staff, equipment, etc.). Typically a government laboratory has areas of deep expertise that attract investment. Another advantage of a government lab is the ease of contracting from a government perspective, since money can be moved faster within the government than from government to the private sector. Informal relationships and marketing the strengths of the labs within the government are important factors in this process.

An important part of leveraging best practices includes understanding lessons learned for growing an S&T workforce. The next part of this paper covers those lessons learned.

PART TWO

Lessons Learned for Growing a Science and Technology Workforce—-DOD Laboratory Leaders

To complement the best practices discussed in Part One, the CTNSP Team developed the following summary of the key elements they believe important for the development of a first class S&T workforce. Again, these concepts are largely based on experiences gained from leading DOD laboratories. May elements, however, are generic and potentially widely applicable. The elements are not listed in a priority order.

- There must be clear and well documented criteria for promotion for scientist and engineers. For example, items that should be considered are:
  - Accomplishments
  - Publications in refereed journals
  - Reference index
  - Honors and awards
  - Leadership skills
  - Peer recognition
- The S&T workforce must maintain a close connection with the "best in class". This is often manifested by being only one phone call away from the best and being active in professional societies, including leadership positions in the societies.
- There must be a strong connection with the ultimate user(s) of the technology being developed. Even the most innovative and productive scientist should be able to articulate the potential benefits to the user when their technology is transitioned.
- The organization must have some level of "patient capital" to allow for long-term investments in promising technologies. It would be desirable for organizations to have a core budget that is relatively stable over the long run. Additionally, the leaders of the core budget should periodically have opportunities to compete for additional funds from their parent or external organizations.
- In cases where external funding is available, the organization must have efficient processes to accept and expend such funding. This is important since many funding organizations have relatively short commitment times for their funding. Strong S&T organizations must have strong business practices.
- A good S&T workforce is made stronger by interactions with other scientist and engineers in similar and related fields. Robust programs to bring other scientists such as Post Docs, National Research Council Fellows, University Professors, or Contractors for tours in an organization are very important. The flexibility offered by such arrangements allows fresh ideas to be periodically introduced to the organization. Additionally, when the other scientists and engineers return to their parent organizations or move on to other jobs, a natural synergy is developed.

- Separate promotion paths for managers and scientist should be identified. Further, it is important for individuals to understand the differences between the two paths and to have opportunities to participate in either path.
- Several of the DOD laboratories have developed innovative compensation programs outside the rigor of the traditional government GS schedule. Examples include pay banding, special hiring authority, and local certification. These programs allow for more competitive salaries and benefits, resulting in better recruitment, improved job satisfaction, and retention.
- It is very important for strong S&T organizations to have appropriate support staffs. A good ratio of trained technicians-to-scientists is of prime importance. Likewise, proper levels of administrative help are important. Highly skilled scientist and engineers become disenchanted when burdened with excessive non-technical issues.
- World class equipment is mandatory to support a world-class S&T workforce. One technique for consideration is to set aside a dedicated portion of budgets for up-grades or purchase of new equipment.
- Opportunities to publish in open literature are important for peer recognition. Clearly for DOD and DHS laboratories, this is not always possible because of security considerations. Nonetheless, the S&T workforce should be provided every effort to publish and to participate in national societies.
- Adequate travel funds to permit broad interactions with peers are necessary. A strong S&T workforce has a "presence" in their technical communities and is recognized for their contributions.
- S&T workforces are strengthened by periodic peer reviews of the quality and relevance of their work. The peer reviewers should be able to share criticisms and accolades with the workforce in a non retribution environment. Additionally, summary reports to organizational leaders are required to help them prioritize workload.
- Periodically hosting "world class" scientists (e.g., Nobel Laureates) to review an organization's work and to interact with the organization's S&T workforce is a very effective morale and quality building technique. Exposing an S&T workforce to the best in their class has many positive elements.

PART THREE

Summary and Recommendations

The study Team of former DOD Laboratory Directors provided lessons learned from their experience and best practices. The lessons learned were presented to the DHS Science and Technology officials (Director and Deputy Director of Research, DHS Director of Office of National Laboratories, DHS Laboratory Directors) at the DHS S&T Headquarters in the Fall of 2010. The common themes of the lessons learned are: a highly competent and dedicated workforce; good connectivity to the larger scientific and technical community; and creating institutional advocacy for the laboratory. In the best practice area of forecasting future work, considering capabilities of other laboratories, and accessing information about competitors, the common thread is having laboratory staff members who are "card-carrying" members in their fields. In addition, the Team responded to the DHS requested areas of evaluation in the areas of balance between short-term versus long-term research, interagency coordination, and quick response to new urgent requirements.

The Team recommends that peer reviews be conducted for the DHS laboratories to evaluate the S&T programs and identify areas of strength and weakness, as well as potential areas of collaboration with other internal or external laboratories in the government, industry, and universities. These reviews can enhance the posture of the laboratories as well as the responsiveness to the customer base. The Team recommends that the DHS labs have strong linkages with the DHS University Centers of Excellence (COEs). Currently, the linkages of the labs to the CoEs are processed through the focal points at the Divisions instead of direct connectivity at the working level. This is a cumbersome process. The Team also recommends that the laboratories conduct a self-evaluation by utilizing the metrics of successful laboratory management identified by the "Federal Advisory Commission on Consolidation and Conversion of Defense Research and Development Laboratories" (See Appendix B). The self-evaluation could be as simple as grading each element with green for positive, yellow for marginal and red for negative score. This evaluation could provide a measure of the general health of the laboratories and reveal the areas that need to be strengthened.

There was general agreement among the former DOD Laboratory Directors that a basic research program was essential to the technical success of their laboratories. Furthermore, a meaningful and in-depth review process of the technical program was necessary. Metrics such as the number of patents, publications, and number of PhDs should be utilized. These and other factors are discussed in detail in a related Defense Technology Paper.[8]

---

[8] Richard Chait, "Perspectives from Former Executives of the DOD Corporate Research Laboratories," Defense and Technology Paper #59, Center for Technology and National Security Policy, March 2008.

The above-mentioned ideas demand successful leadership, which is often based on the four pillars discussed in Appendix C. The CTNSP Team believes that leadership can be learned, and the best way to become a leader is to practice these pillars in both one's professional and personal lives. The members of the CTNSP Team presented their findings on the laboratory management and leadership characteristics to the DHS officials in the October of 2010. The characteristics include: clear and stable missions, a highly competent and dedicated workforce, and stable organization and funding. More details on the characteristics are discussed in Appendix B. The presentation generated extensive dialogue among the laboratory directors.

The DHS laboratories have different management structures, ranging from a Federally Funded Research and Development Center (FFRDC), to Government-Owned/Contractor-Operated (GOCO) models, to Government-Owned/Contractor-Assisted (GOCA) structures. They are also much smaller than the DOD Laboratories in terms of budgets and personnel. As a result, some of these characteristics from DOD experience may not be applicable. The Team initiated direct contacts with the laboratories to discuss best practices and lessons learned. Presentations were made at each individual laboratory either in person or via VTC. Also, an overview of the book *The Splendid Leader* (see Appendix C) was presented to various DHS laboratory leaders and received a positive response. Based on the feedback, a full version of a leadership seminar may be offered to DHS laboratories at a later time.

Appendix A

Center for Technology and National Security Policy (CTNSP) Team Membership

Samuel Musa is a Senior Research Fellow at the Center for Technology and National Security Policy (CTNSP). He was previously Associate Vice President for Strategic Initiatives and Professor of Electrical and Computer Engineering at Northwestern University. Dr. Musa received his PhD in Applied Physics from Harvard University and a BS in Electrical Engineering and a BA from Rutgers University.

William Berry is Deputy Director for Science and Technology and Distinguished Research Fellow at CTNSP. He was previously Acting Deputy Under Secretary of Defense for Laboratories and Basic Sciences. Dr. Berry received his PhD in Zoology from University of Vermont and BS in Biology from Lock Haven University.

Richard Chait is Distinguished Research Fellow at CTNSP. He was previously Chief Scientist, Army Materiel Command, and Director, Army Research and Laboratory Management. Dr. Chait received his PhD in Solid State Science from Syracuse University and a BS degree from Rensselaer Polytechnic Institute.

Timothy Coffee is the former Edison Chair at the Center for Technology and National Security Policy. He was previously the Director of the Naval Research Laboratory. He received his BS degree from the Massachusetts Institute of Technology and an MS and PhD, both in Physics, from the University of Michigan.

John Lyons is a Distinguished Research Fellow at CTNSP. He was previously Director of the Army Research Laboratory and Director of the National Institute of Standards and technology. Dr. Lyons received his PhD from Washington University and a BA degree from Harvard.

Vincent Russo is the former Executive Director of the U.S. Air Force Aeronautical Systems Center. Dr. Russo received his PhD in Metallurgical Engineering from the Ohio State University and BS degree from the University of Rochester.

Donna J. Back is Chief Operating Officer and Chief Financial Officer of Growing Splendid Leaders, LLC. She serves as a leadership instructor for Wright State University's Raj Soin College of Business, the University of Dayton's College of Engineering, and the Ohio State University's College of Engineering. She received her MBA from Wright State University and a BS from Miami University, Ohio.

16

Appendix B

Other Views of Laboratory Characteristics

The characteristics of the best management practices presented below are based on one member's involvement as Director of the Army Research laboratory, and when earlier, as a member of the Federal Advisory Commission on Consolidation and Conversion of Defense Research and Development Laboratories.[9] The Commission's findings were then reported to the Secretary of Defense, in September 1991. The Commission findings included the following criteria for successful lab management:

- Laboratories must have clear and stable missions.
- Laboratories must have highly competent and dedicated workforce.
- Laboratories must have highly qualified and empowered leadership.
- They should have state-of-the-art equipment and facilities.
- Labs should have close relations with the user/customer.
- Laboratories should have a strong basic research component.
- There should be budget stability for the laboratories.
- There should be a champion in senior management above the laboratory.
- There should be strong ties to other laboratories inside and outside the government.

This list generated considerable interest among the DHS Laboratory Directors. In particular, there was interest in applying the characteristics to each laboratory to determine where they stand on a scale of green, yellow and red.

The panel members also presented other characteristics of successful Laboratories as follows:

- Labs should be connected, or be "one phone call away" from world's top experts.
- Use visiting scientists, NRC post docs, etc. for refreshing and new ideas.
- Keep on the cutting edge by maintaining latest state-of-the-art equipment.
- Assure a critical mass for each research topic pursued.
- Assure frequent interaction with customers: co-locate lab staff with customers.
- Obtain support from local community through service on boards, etc.
- Provide opportunities for individual growth of workforce, such as training and advanced degrees.
- Be an effective advocate, not only for technology programs, but for your people. Help them grow!
- The two most important decisions a lab makes are who it hires and who it promotes.

---

[9] http://www.stormingmedia.us/78/7883/A788323.html

- Recognize that scientists and engineers make tradeoffs between real income (salary and benefits) and psychic income (e.g., quality of colleagues, importance/quality of program, quality of facilities, etc.).
- Lab managers and directors should work to create institutional advocacy for the laboratory (within the reporting chain, the policy/political chain, the financial chain, and the scientific and technical community).
- Like most things, it comes down to people, facilities and funding and the calculus that optimizes the lab's output subject to the constraints on the above. As the calculus is done, keep the lab mission in mind.
- The basic metric is: how did the lab do in meeting its internal goals and its external obligations.

The CTNSP Team also utilized material by Dr. Hans Mark, NASA Deputy Administrator between July 10, 1981 and September 1, 1984,[10] who outlined his findings in the book titled *The Management of Research Institutions: A Look at Government Laboratories.*[11] The pertinent selections were presented to the DHS lab directors on December 10, 2009, and include the following:

Successful laboratories are able, again and again, to reinterpret their missions in light of changing conditions—and that is really what the book means by "success."

- **Successful lab should have the following expenditure:** $75,000-$100,000 per employee (1980s dollars).
- **Professional and Support Personnel**: ratio of direct program people to support people is between 1:1 and 1:2. Full-time research to laboratory staff 1:10.
- **Size**: Optimal size is 1,000-7,000 people. If the institutions too small, too little flexibility for a few people to strike out into new territory, or for new ideas to spill over into research work.
- **Mission:** Part of the organic development of the laboratory is that it constantly redefines its own missions and its reason for being.
- **Transfer:** People should transfer in and out of basic research directorate to other units in the course of their careers.
- **Composition:** 10 percent of people in the organization should be devoted to long-term work. Also needed a place for small, non-mission oriented research groups.

Dr. Mark did acknowledge that there was no single approach to hiring and retaining technical personnel, and that an influx of new basic research employees is essential for new ideas. If the laboratory tries to maintain a low average age and an "up or out" policy in promotions, older staff members may indeed feel undervalued and their work will suffer accordingly.

---

[10] http://history.nasa.gov/Biographies/mark.html
[11] http://openlibrary.org/b/OL20701059M/management_of_research_institutions

Dr. Mark proposed in his book that productivity tends to climb after age 50, and that scientific and professional contributions made during this late period are usually more "productive" than "creative." He outlined that following his observations, scientists continue to produce throughout their careers. Lab management can - and should - encourage these people through rotating assignments, offering continuing education and sponsoring some to study at leading management schools or other institutes.

Dr. Mark wrote that all organizations could be divided into two general categories - "mechanistic" and "organismic" entities:
- Mechanistic organizations are hierarchical, with well-established personnel roles and boundaries, and focus on stable conditions.
- An Organismic organization is characterized by a network structure of control and authority, a lateral rather than vertical direction of communications, greater emphasis on commitment to mission than to decisions and judgments of superiors, and acceptance that knowledge may be located anywhere in organization.

According to Dr. Mark, the average "technology development laboratory is a mechanistic organization trying to behave like an organismic one." His book outlines that most successful corporations are organismic organizations where:
- Executives need to push decision-making down to lowest practicable level, with great flexibility as to methods of getting things done.
- Goals are defined and managers partners in a common effort, by delegating authority to work out new tasks in detail.
- Leaders of an institution provide **a vision of the future**.
- Vision must be easy to understand, intellectually challenging, and credible so that it can be turned into reality.
- Many people in the institution must be involved in many different ways to create this vision of the future.

Dr. Mark's findings stated the following principles that should guide executives in shaping their organizations:
- Do not move too far out ahead of the rest of the organization–build on what is already there.
- Personnel development and considerations of the organization's structure cannot be separated.
- Importance of some scientific research in the lab not tied to specific missions.

Appendix C

Leading Technical Organizations

Clearly, the characteristics of a successful laboratory as described above heavily depend on the knowledge, skills, and abilities of the laboratory leaders. Thus, in today's competitive, dynamic, results-driven laboratories, there is undoubtedly a need to assure the highest quality of leadership is present. In the discussions below, the authors present a "framework" that can be used to achieve high quality leadership. This framework for successful leadership can apply to every level in an organization, not just the "senior positions." The framework is identified in the following four pillars:

- Behavior Realities
- Leadership Tenets
- Essence of Leaders
- Life Balance

These Pillars are especially poignant when it comes to leading governmental scientific and engineering institutions at a time of decreasing budgets and changing economic realities.

Pillar One: Behavior Realities

According to a paraphrased quote by Dwight D. Eisenhower, "Leadership is the art of getting people to do what needs to be done because they want to do it."[12] Thus, if one wants to inspire people to do what needs to be done, one should understand why people behave the way they do—a concept envisioned as "Behavior Realities."

Dr. Steve Cato of the Federal Executive Institute points out that Behavior Realities are aligned in three categories: historical, current, and anticipatory.[13] *Historical realities* are those factors that are set early in our lives; such as, cultural background, education, heredity, socialization, personality, values and morals. While these ingrained factors certainly do influence behavior, leaders have a limited ability to affect these factors in others. However, leaders need to be aware of historical realities as they can manifest themselves as behavior responses in the workplace.

The second category of Behavior Realities is *current realities*, which includes such factors as peer pressure, needs, recognition, personal gain, loyalty and habits. These

---

[12] http://leadership.uoregon.edu/resources/quotes
[13] https://www.leadership.opm.gov/

factors are somewhat influential in shaping how employees behave in the workplace. Employees often respond positively to efforts to satisfy their needs for recognition of their accomplishments and contributions. While historical and current realities should not be ignored for their ability to sway behavior, we do not believe they are the dominant reasons for workplace behaviors.

Anticipatory realities, the third category, touch the hearts of people, and therefore, are truly at the core of what most motivates people's behaviors. Anticipatory realities focus on dreams, visions, and "utopian," or far-reaching ideas.

When thinking about Behavior Realities, the most important aspect of understanding why people behave the way they do is this: Anticipatory Realities far outweigh the sum of Historical Realities plus Current Realities in affecting behavior. There is an undeniable link between understanding the power of anticipatory realities and the ability to be a leader. Visions, words, and actions that touch the hearts of others go a long way to impact desired behaviors in the workplace.

Pillar Two: Leadership Tenets

Bookstore shelves are overflowing with theory, advice, and anecdotal experiences on leadership. In this pillar, we have captured our version of the "laws of leadership," or as we refer to them, the Leadership Tenets of successful leaders. These tenets are:

1. Start with the Heart
2. Create Trust
3. Equip People to Excel
4. Use the Word Why
5. Have Fun
6. Cast a Splendid Shadow

We use the term "laws" when explaining these tenets because they are applicable in most leadership situations. Irrespective of where one resides in the organizational pyramid, these laws are relevant.

1. Successful leaders find many ways to *start with the heart*. We have already discussed one such way——the use of anticipatory realities. There are many others including expressing care and concern over non-job related difficulties, expression of appreciation, assigning challenging projects, displaying confidence in an employee's ability to exceed expectations, and tending to the health and safety of employees.

2. As exemplified by newspaper, television, and internet headlines about corporate corruptions, personal misbehaviors, ethical misgivings, and organizational failings, trust has become a scarce commodity in the public and private sectors. Successful leaders *create trust* in the workplace by adhering to the following six beliefs:

- Value Contributions
- Possess Competency
- Constancy of Purpose
- Reciprocal Support
- Create Safe Zones
- Keep Promises

3. The third Leadership Tenet is to *equip people to excel* in achieving the organization's vision and mission while simultaneously reaching their full potential. There are many venues for equipping people to excel such as providing resources, eliminating barriers, investing in their education, mentoring, coaching, and networking.

4. Leaders love to *use the word "why."* They do this because they are compelled to improve their organizations. The intent is to neither challenge nor criticize operational processes and procedures or to put people on the defensive. Instead, the purpose of using the word why is to improve, enhance, and upgrade the status quo as well as to encourage employees to open their minds to new possibilities. Leaders want to leave an organization better than when they joined it—-they want to add value.

5. The next Leadership Tenet, *have fun!,* is often underutilized or even absent from the workplace. Leaders should create an environment where people enjoy coming to work each day, and at the end of the day, they know their efforts and contributions are appreciated. Having fun in the workplace manifests itself in a variety of ways such as celebrations, sporting competitions, social gatherings and cook-offs.

6. The final law of leadership, *cast a splendid shadow,* has to do with leaders living the values, norms, and ethics that represent the desired culture of the organization. Actions of the leader speak volumes about the leader's beliefs, and these beliefs (good and bad) will be reflected back to the leader as behaviors in the workplace.

Pillar Three: Essence of Leaders

Up to this point, the framework of splendid leadership has dealt with the "hard" side of leadership. In pillar three, we will discuss the "soft" side of leadership—-and the "soft" stuff is really the "hard" stuff! The Essence of Leaders is the ability to make feelings work to achieve desired results by using emotions to facilitate exceptional performance from oneself and from others—-and social scientists have documented over a thousand distinct emotions in the workplace! The concept of *essence of leaders* is often referred to as emotional intelligence.

There are five elements of the *Essence of Leaders*:

- Self-Awareness

- Self-Management
- Self-Motivation
- Interpersonal Expertise
- Relationship Building

Great leaders learn the fundamentals of these five elements and practice them daily.

Pillar Four: Life Balance

Until recently, the topic of *Life Balance* was not widely discussed in the workplace. Work and personal life were seen as competing demands. Some people assume that life balance is nothing more than time management. However, life balance is so much more than time management; it is what you **derive** from the things you do with the time you have (Galinsky 1). *Life Balance* underpins the proverbial debate posed by the question: "Can I have it all?" You can have it all; just not all at once. Thus, life balance has to do with work and personal life choices based on priorities.

By definition, *Life Balance* is a personal blend of competing commitments to optimize your full work and life potential over time. It is a process that dissects work and personal activities into mental, physical, emotional and spiritual life accounts. These life accounts are bombarded every day by competing commitments that either make deposits (add to) or withdrawals (subtract from) to the status of your life accounts. These commitments originate from a variety of sources: job, community, relationships, family, health, friends and even oneself. The objective of *Life Balance* is to seek equilibrium among your life accounts over time. It is nearly impossible to maintain optimal balance in all your life accounts at all times. Therefore, through striving to achieve the objective of *Life Balance*, you are constantly making choices among competing commitments that impact the status of your four life accounts.

Leadership Summary

We have outlined the framework of splendid leadership based upon the four pillars: **Behavior Realities, Leadership Tenets, Essence of Leaders** and **Life Balance.** It is important for individuals to not only strive to use these attributes, but to also recognize them in others. We believe that leadership can be learned, and the best way to become a great leader is to practice the four pillars in your professional and personal life. A full treatment of these four Pillars can be found in the book, *The Splendid Leader.*[14]

Selected members of the Team presented the Pillars of leadership to the DHS Laboratories (TSL, NUSTL, Plum Island, National Biodefense Analysis and Countermeasures Center (NBACC) and Chemical Security Analysis Center (CSAC)). The response from the laboratories was very positive, ranging from an invitation to come back with a full day of in-depth presentation and interaction to a presentation on selected areas of interest to the laboratory. Some of the laboratory leadership felt that the presentation should focus more on building trust among the personnel in light of recent reductions in force. Some of the laboratories have young staff and there is a sense of growing pains that need to be addressed. Follow-up activities are anticipated with each laboratory.

Works Cited

Boyatzis, Richard, Daniel Goleman, and Annie McKee. *Primal Leadership.* Boston, MA; Harvard Business School Press, 2004.

Cato, Steve. The Federal Executive Institute, Executive Development Course, Currently Distinguished Honorary Fellow, Antioch University, Seattle, WA, 1990.

Fitz, Raymond, SM. Notes from personal interview, 2006.

Galinsky, Ellen. "Dual Centric—A New Concept of Work Life." Families and Work Institute, 2004.

Russo, Vincent J., and Donna J. Back. *The Splendid Leader.* Dayton, OH; BookFactory Publishing, 2008.

---

[14] "The Splendid Leader," Vincent J. Russo and Donna J. Back. Dayton, OH; BookFactory Publishing, 2008

Appendix D

Introduction to DHS Laboratories

A brief description of the DHS Laboratories and their structures is presented here to set the stage for the potential application. DHS Science and Technology Directorate oversees five national laboratories. **National Biodefense Analysis and Countermeasures Center (NBACC)** provides the nation with the scientific basis for awareness of biological threats and bioforensic analysis to support attribution of their use against the American public.[15] It is operated as a Federally Funded Research and Development Center (FFRDC) for the Department by the Battelle National Biodefense Institute, LLC. The Center is programmatically aligned to the Office of National Labs under the Director of Research and Development Partnerships. Its activities include studies and research to better understand current and future biological threats; assess vulnerabilities and conduct risk assessments; determine potential impacts to guide countermeasure development and conduct bioforensics analysis of evidence from a biocrime or terrorist attack to help in the investigation of the perpetrators and method of attack. NBACC is located within the National Interagency Biodefense Campus at Fort Detrick, Maryland.

National Urban Security Technology Laboratory (NUSTL) tests, evaluates and analyzes homeland security technologies and capabilities while serving as a technical authority to first responder, state and local entities in protecting our cities.[16] NUSTL also serves as a Federal technical authority promoting the successful development and integration of homeland security technologies into operational end-user environments. It is a government-owned/government-operated facility organized under and operated by the Department's Science and Technology Directorate. The laboratory is programmatically aligned to the Director of Support to the Homeland Security Enterprise and First Responders. The Laboratory is located in New York City.

Transportation Security Laboratory (TSL) is tasked with independent testing and validation of technology; maintaining "Gold Standard" electronic databases for data validation. Additionally, it develops RDT&E solutions focused on improvised explosive devices (IEDs) and other potential threats to the nation's transportation systems.[17] The Laboratory is a government-owned/ government-operated facility organized under and operated by the Department's Science and Technology Directorate. The laboratory is programmatically aligned to the Test and Evaluation & Standards Office under the Director of Acquisition Support and Operations Analysis. It is located at the William J. Hughes Technical Center, Atlantic City International Airport, New Jersey.

Chemical Security Analysis Center (CSAC) provides DHS S&T with the scientific basis for the awareness of chemical threats and the attribution of their use against the

---

[15] http://www.dhs.gov/files/labs/scitech.shtm
[16] Ibid.
[17] Ibid.

American public.[18] It analyzes and integrates chemical threat characterization data, including toxic industrial chemicals, chemical warfare agents, and other chemicals of interest. The Center is programmatically aligned to the Office of National Labs under the Director of Research and Development Partnerships. The Laboratory is staffed with a few government employees and several contractors. It is located at the Edgewood Area of Aberdeen Proving Ground and draws upon the expertise in chemical defense, chemical agents, and toxic industrial chemicals resident at the Proving Ground.

Plum Island Animal Disease Center (PIADC) protects the nation against animal diseases that could accidentally or deliberately be introduced into the country, for example foot-and-mouth disease (FMD).[19] It works to provide confirmatory diagnostic capability for specific high-consequence foreign animal diseases in livestock, such as cattle, sheep, and swine. The Laboratory is programmatically aligned to the Office of National Labs under the Director of Research and Development Partnerships. It is located in Plum Island, New York. The Center will be relocated to Manhattan, Kansas as part of the **National Bio and Agro-Defense Facility** (NBAF), which is tasked with protecting the nation's agriculture livestock and public health against numerous foreign animal and emerging zoonotic diseases.[20] The NBAF will be built on a site on Kansas State University adjacent to the existing Biosecurity Research Institute.

The customer of the study was the Director/Deputy Director for Research, Science & Technology Directorate at the Department of Homeland Security. The Director of Research was one of three functional portfolio managers reporting to the Undersecretary of Science and Technology. The others were the Directors of Innovation and Transition. A new organization of the Office of the Undersecretary for S&T was completed in October 2010 with four Directors and a Chief Scientist. The Directors are the previously mentioned Director of Research and Development Partnerships, Director of Acquisition Support and Operations Analysis, Director of Support to the Homeland Security Enterprise and First Responders, and Director of Homeland Security Advanced Research Projects Agency. There are six Divisions that report to the latter Director where the majority of the programmatic resourcing and execution originates. The lion's share of the DHS laboratories research budget funnels through one of the six Divisions within S&T.

---

[18] Ibid.
[19] Ibid.
[20] Ibid.

www.ingramcontent.com/pod-product-compliance
Lightning Source LLC
Chambersburg PA
CBHW081420170526
45166CB00010B/3415